TODAY'S BIGGEST ANIMALS

TODAY'S BIGGEST ANIMALS

By Kathleen N. Daly

Pictures by
Tim and Greg Hildebrandt

Golden Press • New York
Western Publishing Company, Inc.
Racine, Wisconsin

Library of Congress Catalog Card Number: 76-27611

Once dinosaurs that were much bigger than elephants lived on earth. Then the dinosaurs disappeared, but there are still giants among today's animals.

BLUE WHALE

The blue whale is the biggest animal in the world. It is even bigger than any dinosaur that ever lived. It may weigh 150 tons and be more than 100 feet long.

This giant whale lives in the ocean. But it is a mammal, not a fish. Like all mammals, the mother whale feeds her baby with milk from her body. The baby calf is about 25 feet long when it is born. The mother swims under the newborn baby to raise its head above water so it can breathe.

Whales breathe air, just as we do. They have to come to the surface to do this. When the whale breathes out through the blowholes at the top of its head, the moist air forms a fountain of spray.

ELEPHANT

The biggest land animal is the African elephant. A large male weighs more than 12,000 pounds and is more than 10 feet tall. (The Indian elephant is smaller, with shorter tusks.)

In Africa, elephants travel in herds, searching for food. They use their trunks to reach high into trees for leaves and twigs. They use their tusks to dig for roots in the ground.

At the waterhole, elephants cool off by filling their trunks with water and showering themselves.

Baby elephants travel with the herd when they are only a few days old. The mothers and the other grownups protect the babies from danger.

Sometimes the hippo baby rides on its mother's back in the water.

HIPPOPOTAMUS

Next to the elephant, the hippopotamus is the biggest animal. Its name means "river horse." This bulky beast is 14 feet long and weighs up to 8,000 pounds.

In the warm, muddy rivers where they live, hippos are graceful swimmers. They spend most of the day in the water because the sun is too hot for their tender skins. At night, they leave the river to feed on tall grasses.

Often a hippo just stands in the water with only its nose and puffy little eyes showing. It has a very sleepy look. When a hippo gives a big yawn, its huge pink mouth opens like a door.

RHINOCEROS

There are different kinds of rhinoceroses. Some have one horn. Some have two horns. Some are not as big as others, or are a different color. The biggest rhinos weigh up to 7,000 pounds and are more than five feet tall.

One kind of rhino has a leathery skin with thick, armorlike plates. It looks like an armored tank as it lumbers across the prairie.

With its bulky body and short legs, the rhino doesn't look as if it could run very fast. But it can. If it hears a strange sound or picks up a strange smell, it will often lower its head and charge. When a big rhino charges, the earth trembles.

GIRAFFE

The tallest of all animals is the giraffe. Some giraffes are more than 18 feet tall. Even newborn baby giraffes are as tall as a man.

With such a long neck, it is easy for a giraffe to feed on leaves and twigs high in the treetops. But when a giraffe is thirsty its long neck can be a problem. It has to spread its front legs far apart to lower its head to the water.

Giraffes have big, velvety eyes shaded by thick lashes. Because they are so tall, they can see much farther than other animals and can give warning of danger.

KODIAK BEAR

When the big, brown Kodiak bear stands up, it is nearly 11 feet tall. It weighs a lot, too—about 1,500 pounds. It has huge feet and long, dangerous claws. The Kodiak is the biggest bear, bigger even than the giant grizzly.

In the winter, two tiny bear cubs are born in a cave. They stay in the cave until spring comes. Then their mother teaches them where to find honey and berries, and how to catch fish. When winter comes again, the bears return to the cave, where they sleep away the long, cold months.

The Kodiak gets its name from the island off the Alaskan coast where it lives.

MOOSE

Another great animal of the northern forests is the moose. It is the largest member of the deer family.

The male moose grows horns, called antlers. Each year these grow bigger. Sometimes they are more than six feet across and may weigh over 100 pounds. The huge antlers are shed every winter but grow back in the summer.

A big moose is seven feet high at the shoulder and may weigh up to 1,800 pounds. When he comes crashing through the forest, the other animals take cover.

TIGER

The tiger, with its handsome striped coat, is one of the most beautiful of all animals. When it moves through the tall, waving grasslands, its stripes keep it almost hidden.

Tigers rest during the day and hunt in the evening shadows. Even though they weigh 500 pounds, these big animals make hardly a sound as they creep up on their prey. They are patient hunters. They will even swim across rivers in search of food.

Most tigers are about the same size as lions. But the male Siberian tiger may be bigger than any lion. It lives in the far North, where winters are cold and snowy. Few people have seen this biggest of cats, for it is very rare.

LION

Lions are the only big cats that live together in family groups. These families are called "prides," and sometimes there are as many as 20 lions in one pride. The spotted cubs play together like kittens, while the fathers nap in the heat of the day and the mothers keep a sharp lookout.

When the cubs are old enough, their mothers take them along on hunting trips. The cubs learn how to hide in the tall grass and how to run and pounce like mighty hunters.

OSTRICH

The eight-foot-tall ostrich is the biggest bird in the world. This strange-looking creature has a small head, a long, bare neck, and long, bare legs. Beautiful feathers cover the rest of its body. The ostrich cannot fly, but it can run very fast on its powerful legs. Its legs can also kick out dangerously at enemies.

Ostrich eggs are about the size of a child's football. Out of them hatch chicks with spiky feathers.

CONDOR

One of the biggest of all flying birds is the condor. When its wings are spread in flight they are nearly 10 feet across. These birds live in the mountains, where they glide on air currents, searching for small animals far below.

TRUMPETER SWAN

The trumpeter swan is one of the largest water birds. This graceful swan "trumpets" a loud call that can be heard for miles.

KOMODO DRAGON

Many lizards are small and good to have around because they catch lots of insects. But the lizard that lives on Komodo Island is big enough to be scary. It is 10 feet long and weighs more than 300 pounds. A lash of its heavy tail can knock over even a large animal.

The Komodo dragon prowls the island, hunting for deer to eat. When it finds one, the terrible dragon attacks it, biting down hard with razor-sharp teeth and slashing with its long claws.

CROCODILE

The biggest crocodiles are more than 20 feet long. Like all reptiles, they spend most of their time basking in the warm sun or finding a shady spot in which to cool off. In a river, they often lie just below the water. They look like floating logs, except for their watchful eyes.

The great, smiling jaws of a crocodile are full of sharp teeth, and its powerful tail is used like a whip. Other animals stay away when this hungry reptile is looking for a meal.

Young crocodiles hatch from eggs. For the first few days the babies follow their mother around, but then they go off on their own.

ALLIGATOR

Alligators are not as big as crocodiles, but they are pretty big just the same. They look a lot like crocodiles. But when an alligator's mouth is closed, none of its teeth show. That's one way to tell the two reptiles apart. Also, an alligator's head is shorter and thicker than a crocodile's.

ELEPHANT SEAL

The elephant seal gets its name from its droopy nose that reminds people of an elephant's trunk. Male elephant seals are much bigger than the females. A big male is more than 20 feet long and weighs 5,000 pounds.

Hundreds of elephant seals gather on rocky beaches in spring, when their babies are born. The father seals protect the nurseries, snorting loudly through their "trunks" to keep other animals away.

Baby seals are called pups. They are born with black, woolly coats. When the babies are old enough, their mothers teach them to swim.

LEATHERBACK SEA TURTLE

Most turtles have hard, bony shells. But not the leatherback sea turtle. Its shell is covered with tough leather plates.

This monster of the seas can weigh up to 2,000 pounds and is about nine feet long. The mother turtle only comes ashore when it is time to lay her eggs. She digs a hole in the sand, leaves the eggs, and returns to the sea.

The baby leatherbacks seem to know that their home is in the sea, and they hurry across the sand to reach it as soon as they hatch. The tiny babies are good swimmers, using their fat little flippers to push them through the water.

GORILLA

Gorillas are the biggest apes. They weigh as much as 450 pounds. But when a baby gorilla is born it weighs about four pounds. (A human baby weighs about seven pounds.) The mother gorilla takes good care of her baby and carries it everywhere with her. When the baby is older, it likes to ride around on its mother's back.

Gorillas live in family groups. They travel most of the day, looking for fruits and plants to eat. At night, they make large, leafy nests in trees and sleep late into the morning.

ANACONDA

The anaconda is the biggest and heaviest snake in the world. When it stretches out to its full size, it may be over 30 feet long, and may weigh 600 pounds.

This giant snake lives in swamps. In the daytime, it usually lies in the water, with only its eyes showing. At night, it hunts for a bird or other small animal. When it finds one, it coils its great body around the victim and crushes it.

KING COBRA

Some snakes kill their prey by biting them with their poisonous fangs. The biggest of these poisonous snakes is the king cobra. It mostly eats other snakes.

Just before the cobra attacks, its head swells up to many times its normal size. It looks very, very fierce.

WHALE SHARK

There are many big fish in the sea. But the biggest of them all is the whale shark. It may be 60 feet long and weigh about 30,000 pounds. It is even bigger than many whales.

The whale shark has a huge mouth. Two big people could fit inside it. And it has 3,600 or more teeth! But this giant of the seas does not attack people. It eats only tiny fish and plants. Of course, such a big shark needs to eat thousands of little fish each day.

DEVILFISH

The devilfish is another ocean giant. It measures 20 feet across and looks as if it has great flat wings. It flaps these "wings" as it swims, and sometimes leaps high out of the water.

GRAY WOLF

Many fearsome tales are told about the great gray wolf of north. But this largest member of the wild dog family, like most wild animals, stays away from people and will only attack in self-defense.

The gray wolf is about five feet long and weighs about 110 pounds. (A dachshund weighs about 25 pounds.)

Wolves are loyal to their family group and they stay together for many years. Like dogs, wolves are very intelligent. They hunt in packs. Sometimes they kill cattle or sheep, and they in turn are killed by farmers and hunters. Only a few of these big dogs remain. Conservationists must protect them, and all endangered animals, if any are to remain in the world.